Arthur Gamgee

Researches on the Blood

On the action of nitrites on blood

Arthur Gamgee

Researches on the Blood
On the action of nitrites on blood

ISBN/EAN: 9783337391249

Printed in Europe, USA, Canada, Australia, Japan

Cover: Foto ©berggeist007 / pixelio.de

More available books at **www.hansebooks.com**

XXIV. *Researches on the Blood.—On the Action of Nitrites on Blood.* By ARTHUR GAMGEE, *M.D., F.R.S.E., Assistant to the Professor of Medical Jurisprudence in the University of Edinburgh.* Communicated by Professor FRANKLAND, *F.R.S.*

Received April 1,—Read May 7, 1868.

Introductory Propositions[1].

IN the following propositions the leading facts which have been ascertained by previous observers with regard to the colouring-matter of the blood and its relation to gases have been condensed.

1. The colouring-matter of blood, to which the names of Cruorine[2] or Hæmoglobin[3] have been given, occurs in solution in the blood-corpuscles, but may by suitable treatment. be obtained from them in the form of crystals, which when seen individually are of a yellow, and when seen collectively are of a reddish colour. Their solution possesses the property of absorbing light so as to yield a remarkable spectrum characterized by two very well-defined absorption-bands situated in the yellow and green portions of the spectrum, and by the cutting off of the greater part of the more refrangible rays[4].

These crystals are identical with those noticed by LEIDIG[5], REICHERT[6], and KÖLLIKER[7], and afterwards more fully described by FUNKE[8], KUNDE[9], and LEHMANN[10].

2. The composition of these crystals appears to be perfectly definite. The following are the results of the analyses of two independent observers, C. SCHMIDT[11] and HOPPE-SEYLER[12]:—

[1] The author has thought it right to introduce these propositions in order to remove any doubts which might exist in the mind of the reader with regard to the identity of Cruorine or Hæmoglobin with the blood-crystals of FUNKE and LEHMANN, and to place in as clear a light as possible the leading facts which have been made out on this subject.

[2] "On the Reduction and Oxidation of the Colouring-matter of the Blood, by Professor STOKES, F.R.S.," Proceedings of the Royal Society of London, vol. xiii. p. 357, paragraph 8.

[3] BERZELIUS applied the term Hæmatoglobulin to the colouring-matter of blood. HOPPE-SEYLER and other German physiologists have adopted the term Hæmoglobin.

[4] HOPPE, VIRCHOW's Archiv. vol. xxiii. p. 446 (1862).

[5] Zeitschrift f. wiss. Zoologie. 1849, Bd. I. p. 116.

[6] MÜLLER's Archiv, 1849, p. 197.

[7] Zeitschrift f. wiss. Zoologie, 1849, Bd. I. p. 261.

[8] Zeitschrift f. rat. Med. Bd. I. p. 172 (1851).

[9] Zeitschrift f. rat. Med. Bd. II. p. 271.

[10] Qualis sit Hæmocrystallines natura chemica. Programm. Jena, 1856 (quoted by HOPPE). LEHMANN's 'Physiological Chemistry,' translated by Dr. DAY, Cavendish Society, vol. iii. pp. 485–495 (1854).

[11] Ueber Blutkrystalle. Dorpat, 1862, p. 33.

[12] Medicinisch-chemische Untersuchungen, zweites Heft, p. 187.

H. S.	C. S.
C = 53·64	C = 53·85
H = 7·11	H = 7·32
N = 16·19	N = 16·17
S = 0·66	S = 0·39
Fe = 0·43	Fe = 0·43
O = 21·02	O = 21·84

3. The colouring-matter exists in blood in combination with oxygen; this oxygen cannot be discovered by pyrogallic acid[1], nor is its absorption by the colouring-matter regulated by DALTON and HENRY's law of absorption[2]. The combination is, however, of a loose character, and is broken up by the addition of reducing-solutions to blood[3], as well as by boiling blood, or merely exposing it in a Toricellian vacuum[4]. The blood-colouring-matter freed from its loose oxygen differs from the oxidized substance, (a) in colour[5], (b) in its absorption-spectrum[6]. To this substance the names of reduced or purple cruorine[7] and reduced hæmoglobin[8] have been given.

4. By passing hydrogen, carbonic acid, or laughing-gas through blood the oxygen is expelled, i. e. the colouring-matter is reduced[9]. Blood which has been reduced by hydrogen or carbonic acid may be made to furnish crystals of reduced hæmoglobin, which are isomorphous with those of the oxidized body, but which differ from it in colour and in optical characters when examined with the spectroscope[10].

5. " We may infer from the facts above mentioned that the colouring-matter of blood, like indigo, is capable of existing in two states of oxidation, distinguishable by a difference of colour and a fundamental difference in the action on the spectrum. It may be made to pass from the more to the less oxidized state by the action of suitable reducing agents, and recovers its oxygen by absorption from the air" (STOKES)[11].

6. The loosely combined oxygen of the colouring-matter may be expelled by carbonic oxide gas[12]; one volume of carbonic oxide takes the place of one volume of oxygen[13]. The compound thus formed differs from the O-compound in not yielding a less oxygenized substance when treated with suitable reducing agents[14]. It possesses an absorp-

[1] BERNARD, Propriétés des liquides de l'organisme, t. i. p. 337 (1859).

[2] LOTHAR MEYER, " Die gase des Blutes," Zeitschrift f. rat. Med. Bd. VIII. p. 257 (1857).

[3] STOKES, op. cit.

[4] HOPPE-SEYLER, Med.-chemisch. Untersuchungen, zweites Heft, p. 191.

[5] STOKES, Proceedings of Royal Society, loc. cit.   [6] Ibid.   [7] Ibid.

[8] KUHNE, Lehrbuch der Physiologischen Chemie, 1866, zweites Lieferung, p. 218.

[9] KUHNE, op. cit., zw. Lief. p. 210.   L. HERMANN, " Ueber die wirkungen des Stickoxydgases auf das Blut,"
MÜLLER's Archiv, 1865, p. 470.   [10] KUHNE, op. cit. p. 218.

[11] STOKES, Proceedings of the Royal Society, vol. xiii. p. 357, paragraph 8.

[12] CLAUDE BERNARD, Leçons sur les effets des substances toxiques et médicamenteuses. Paris, 1857, p. 158.
Propriétés des liq. de l'organisme, vol. i. p. 365.

[13] De sanguine oxydo carbonico infecto. Auctor LOTHAR MEYER, Diss. Inaug. Chym.   Vratislaviæ 1858·
" Ueber die Einwirkung des Kohlenoxydgas auf Blut," Zeitschrift f. rat. Med. Dritte Reihe. V. Band. 1859, p. 82.

[14] HOPPE-SEYLER, Zeitschrift für Anal. Chemie, vol. iii. p. 439.

tion spectrum almost identical with that of the O-compound[1]. It may be obtained in well-defined crystals which are isomorphous with those of the O-compound[2].

7. Similarly the CO-compound may be decomposed by nitric oxide, one volume of CO being apparently replaced by one volume of N_2O_2. The body thus formed may be obtained in crystals isomorphous with those of the O- and the CO-compounds. Like the latter it is unacted upon by reducing solutions[3].

8. Hydrocyanic acid when added to blood appears to combine with the colouring-matter; the nature of the compound formed has not, however, been ascertained. It may, however, be obtained in crystals identical with those of the O- and CO-compound. Its spectrum is identical with the former and, like it, is capable of reduction[4].

9. When heated, or when treated with strong acids and alkalies, hæmoglobin furnishes amongst other products of decomposition, an insoluble body called hæmatin[5], which was until recently supposed to be the colouring-matter of blood[6]. This body possesses no power of combining with oxygen. The absorption-coefficient of oxygen for a solution of hæmatin is nearly the same as for water[7].

Note.—The author has purposely avoided entering upon the description of the interesting physical properties and chemical reactions and relations of hæmatin, as these do not immediately concern the object of this memoir.

On the Action of Nitrites on Blood.

No one has hitherto investigated the action which the nitrites exert upon the blood and upon its colouring-matter. Two of the oxides of nitrogen have been studied in their relation to blood by Dr. LUDIMAR HERMANN, viz. laughing-gas[8] and nitric oxide[9]. With regard to the first of these, it has been shown that it acts upon blood and hæmoglobin very much as hydrogen, carbonic acid, and nitrogen, whilst in reference to the second, HERMANN has shown that it actually combines with the colouring-matter.

My attention was directed to the peculiar action of nitrites upon the blood-colouring-matter by observing that the blood of mice poisoned by exposure to an atmosphere impregnated with the vapour of nitrite of amyl presented a chocolate-colour. I was thus

[1] HOPPE-SEYLER, Medicinisch-chemische Untersuchungen, zweites Heft, p. 203.

[2] HOPPE-SEYLER, Med.-chem. Unters. p. 201.

[3] Dr. LUDIMAR HERMANN, "Ueber die wirkungen des Stickstoffoxydgas auf das Blut," MÜLLER's Archiv, 1865, pp. 469–481.

[4] HOPPE-SEYLER, Med.-chem. Untersuchungen, 11 Heft, p. 206. PREYER, VIRCHOW's Archiv, 1867, Sept. p. 125.

[5] HOPPE, VIRCHOW's Archiv, Bd. XXIII. p. 446. STOKES, *op. cit.* p. 355 *et seq.*

[6] The reader who wishes to become acquainted with the different methods employed formerly for separating hæmatin are referred to GORUP-BESANEZ, Lehrbuch der Phys. Chemie, Braunschweig, 1862, p. 168 *et seq.* BERZELIUS clearly distinguished the difference between the unaltered colouring-matter and the coagulated colouring-matter.—Traité de Chimie, par J. T. BERZELIUS, Paris, 1833. t. vii. pp. 48–65.

[7] LOTHAR MEYER, De sanguine oxydo carbonico infecto, pp. 9–12.

[8] " Ueber die physiologischen wirkungen des Stickstoffoxydulgases," MÜLLER's Archiv, 1864, p. 521.

[9] " Ueber die wirkungen des Stickstoffoxydgas auf das Blut," MÜLLER's Archiv, 1865, p. 469.

led to find that the optical properties of blood were considerably modified by the action of nitrites. I embodied the results of my first observations in a preliminary note which I read before the Royal Society of Edinburgh on the 4th of March, 1867[1]. At that time I was acquainted with little more than the changes in the optical characters induced by nitrites.

The nature of the changes and the influence which they exert upon the relation of the blood to gases are chiefly to be discussed in the present paper.

I. *On the Changes in the Colour of the Blood induced by Nitrites.*

When defibrinated and well-arterialized blood is mixed with a solution of nitrite of potassium or nitrite of sodium, its colour becomes almost immediately altered ; from its beautiful florid-red it changes to a more or less chocolate-brown colour. The rapidity with which this change takes place appears to vary considerably in the case of different samples of blood. Just as great differences are found to exist in the relative rate with which the colour of blood (and its spectrum) is affected by standard solutions of alkalies and acids, so considerable variations occur in the rapidity of the action of nitrites. In the course of this investigation the nitrites whose action on blood has been chiefly examined have been those of potassium, sodium, silver, and amyl. The alkaline nitrites were always prepared by decomposing repeatedly crystallized nitrite of silver by means of an exactly equivalent quantity of the alkaline chloride. The nitrite of amyl was prepared by the action of nitrous acid upon repeatedly fractionated amylic alcohol. The product was also repeatedly fractionated, and only those portions which boiled between 95° and 100° C. collected as pure.

The differences in the rapidity with which nitrites and other active substances act upon blood, probably depend much upon physical conditions affecting the blood-corpuscles. The freshly-drawn blood of the Dog is almost instantaneously affected in colour when mixed with solutions of nitrites, whereas in the case of the blood of the Ox and Sheep occasionally twenty minutes, or even longer, will elapse before a solution of nitrite of potassium or sodium effects the characteristic change. The blood of the Dog, it must be remarked, is used in preference to that of the Ox and Sheep in the preparation of hæmoglobin, because its blood-corpuscles when treated with water burst so much more readily than those of the Ox and Sheep.

When it is desired to act upon blood with nitrite of amyl, the latter should be dissolved in alcohol, and a few drops of the alcoholic solution added to the blood.

Solutions of hæmoglobin are immediately affected by the addition of nitrites.

When blood has, through the addition of nitrites, assumed a chocolate-colour, the latter is immediately changed to a red on the addition of ammonia.

The change in colour induced by nitrites in no way depends upon an alteration in the shape of the blood-corpuscles, as I have ascertained by numerous experiments.

[1] "Note on the Action of Nitric Oxide, Nitrous Acid and Nitrites on Hæmoglobin," Proceedings of the Royal Society of Edinburgh, 1867, No. 73, p. 108.

II. *On the Spectrum of Blood acted upon by Nitrites.*

If normal arterialized blood be suitably diluted so as to admit of an accurate examination of its absorption-spectrum, and if after the position of the two characteristic absorption-bands has been noted, a solution of any nitrite be added to it, it will be noticed that when the solution begins to change from a red to a brown tint, the absorption-bands undergo notable changes. The two sharply-defined absorption-bands of the oxidized colouring-matter become fainter and fainter, and are, indeed, only visible when a comparatively thick layer of the fluid is examined. At the same time, if the layer be sufficiently thick, it will be noticed that an additional, though comparatively faint, absorption-band appears in the red : this band appears absolutely to coincide with that of acid hæmatin ; it is seen to greatest advantage when so thick a layer of solution is examined that all but the red rays are cut off. The complete change, as seen with the spectroscope, is always coincident with the complete change of colour. If, now, the solution be made alkaline by the addition of ammonia, the colour changes from the chocolate-brown to blood-red again ; simultaneously the absorption-band in the red disappears, and the two absorption-bands between D and E become more distinct again. In addition, however, it is noticed that the portion of the spectrum occupying the confines of the yellow and orange has become shaded by a less well-defined absorption-band.

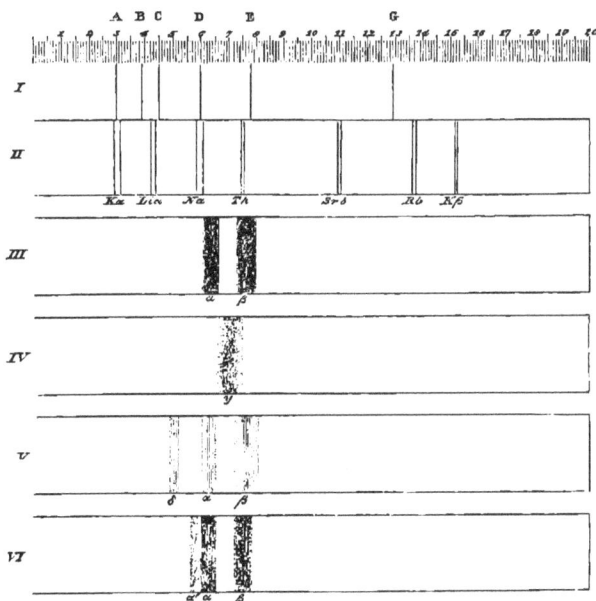

I. indicates approximately the position of FRAUNHOFER's lines A, B, C, D, E, and G in relation to the millimetre scale of the spectroscope used in this research.

4 M 2

II. indicates the position occupied by several of the bright lines of the metals in relation to the millimetre scale of the spectroscope used.

$$K\alpha \ . \ . \ . \ . \ . \ . \ 30$$
$$Li\alpha \ . \ . \ . \ . \ . \ . \ 43$$
$$Na \ . \ . \ . \ . \ . \ . \ 60$$
$$Th \ . \ . \ . \ . \ . \ . \ 75$$
$$Ir\delta \ . \ . \ . \ . \ . \ . \ 110$$
$$Rb\beta \ . \ . \ . \ . \ . \ 136$$
$$Rb\alpha . \ . \ . \ . \ . \ . \ 137$$
$$R\beta \ . \ . \ . \ . \ . \ 152$$

III. indicates the absorption-spectrum of oxidized hæmoglobin as seen in relation to the millimetre scale of my instrument.

$$\alpha \ . \ . \ . \ 60\text{--}65$$
$$\beta \ . \ . \ . \ 72\text{--}78.$$

IV. indicates the absorption-band of reduced hæmoglobin (purple cruorine).

$$\gamma \ . \ . \ . \ 65\text{--}74.$$

V. indicates the spectrum of blood which has been treated with nitrites. The crystals of the nitrite compound of hæmoglobin also exhibit the same spectrum.

$$\alpha \ . \ . \ . \ 60\text{--}65 \ \text{(very faint)}.$$
$$\beta \ . \ . \ . \ 72\text{--}78 \ \text{(very faint)}.$$
$$\delta \ . \ . \ . \ 49\text{--}51.$$

VI. indicates the absorption-bands of the compounds of nitrite with hæmoglobin after the addition of ammonia.

$$\alpha' \ 56\text{--}59.$$
$$\left.\begin{array}{l} \beta \ \ 60\text{--}65 \\ \alpha \ \ 72\text{--}78 \end{array}\right\} \text{blacker and more sharply defined than in V., not so much so as in III.}$$

The spectroscope which I chiefly employed in this research was one with a single prism, constructed by DESAGA, of Heidelberg. I ascertained exactly the positions of the most characteristic lines of the metals in relation to the millimetre scale of the instrument. The position of the absorption-bands described by me will be better seen by comparison with the drawing, in which the position of the bright lines of the metals is marked. I have also, for comparison, inserted spectra of oxidized and reduced cruorine, in which the position of the absorption-bands of these substances is seen in relation to the millimetre scale of my instrument. In the case of the oxidized colouring-matter, I shall distinguish the absorption-bands in the yellow and green as α and β respectively; that of reduced cruorine I shall refer to as γ.

Obs. I. A solution of defibrinated Ox's blood was made by diluting 5 cub. centims. to the volume of one litre with distilled water. The solution was placed for examination in a hollow glass prism, which permitted me to examine definite and varying thicknesses of fluid.   The prism had a capacity of 200 cub. centims.

With a layer of solution 2 centimetres thick, the following observations were made :—

Extent of spectrum visible . . .   30–110
Absorption-band in yellow (α)   .   60–65
Absorption-band in green (β) . .   71–76

Five drops of nitrite of amyl were now added ; after ten minutes a slight fading of the band in the green and a brown tinge in the colour of the fluid was noticed.   After fifteen minutes the two absorption-bands α and β had become so exceedingly faint and ill-defined as not to be capable of accurate measurement.   There was also a faint absorption-band in the red, having its centre at 49.   A stratum of the fluid 3·8 centimetres broad was now examined with the following results:—

Extent of spectrum visible . . .   35–110
Absorption-band α . . . . . .   60–65 (very faint)
Absorption-band β . . . . . .   71–76 (very faint)
Absorption-band δ (in red) . . .   48–51

The fluid was now made faintly ammoniacal.   The band in the red instantly disappeared ; the hæmoglobin bands became much darker ; the orange was, however, shaded partly by a faint absorption-band which appeared to run into the band α.   The following observations were made :—

Length of visible spectrum . . .   30–110
Absorption-band α   60–65
Absorption-band α'   56–59
Absorption-band β   72–78

The observation just detailed is merely taken as an example of what always occurs when a nitrite acts upon blood, and when the latter is afterwards treated with ammonia.

Obs. II. If after the action of a nitrite upon blood and the subsequent addition of ammonia, a solution of sulphide of ammonium or a reducing iron-solution be added, there is a very rapid and extraordinary change, viz. the spectrum of the original blood is first re-established, and thereafter gives place to that of reduced hæmoglobin.

The reduced blood then differs in no respect from blood that has been reduced without the previous addition of nitrites; it yields, when shaken with air, the spectrum of the oxidized colouring-matter, and instead of a brown chocolate-colour, it possesses again the colour of blood.   The action of the reducing-solution in first of all changing the nitrite spectrum to that of oxidized hæmoglobin takes place quite independently of the external air.   If, for example, the blood-solution in which a nitrite had induced its characteristic action be placed in the prism bottle, and then a long pipette full of sulphide of ammonium be plunged to the bottom of it, the two sharply-defined absorption-bands of O-Hb at once appear at the point where the fluids have come in contact; whereas it is only after some time that the further reducing-action of the sulphide makes itself manifest.

Obs. III. Five cub. centims. of freshly defibrinated blood of the Ox were diluted to the volume of one litre, and the solution was then examined in a hollow prism. A layer of solution two centimetres thick was examined. The spectrum of this solution was as follows :—

$$
\begin{aligned}
&\text{Band } \alpha \quad . \quad . \quad . \quad . \quad . \quad . \quad 60\text{--}65 \\
&\text{Band } \beta \quad . \quad . \quad . \quad . \quad . \quad . \quad 72\text{--}76 \\
&\text{Extent of spectrum} \quad . \quad . \quad . \quad 30\text{--}110
\end{aligned}
$$

One cub. centim. of a solution of nitrite of potassium of unknown strength was added to 200 cub. centims. of the above blood-solution. Almost instantaneously the bands became much fainter. About two minutes after the addition of the nitrite, on examining a stratum two centimetres broad, the two absorption-bands had become almost invisible, appearing as mere shadings of the spectrum. There was a very faint band with its centre at 50. On examining thicker strata of fluid the original bands could be seen more distinctly, and the band in the red was very well marked. A solution of ammonia was added to the fluid. The red band disappeared at once, and the two bands became more intense. The orange was observed to be shaded, a faint absorption-band appearing to overshadow it and to join the band α; as in other observations to determine this point it was impossible for me to decide whether the band in the orange was separated from the band α by any unshaded portion, i. e. whether it was an absolutely separate band. On bringing a soda-salt into the flame of the lamp whose light was passed through the blood-solution and examined, the yellow line was seen to be bounded above by a very marked absorption-band extending from 60 to 65, and below by the shadowy but distinct band in the orange (α'). The following observations were now made :—

$$
\begin{aligned}
\alpha \quad &. \quad . \quad 60\text{--}65 \\
\alpha' \quad &. \quad . \quad 56\text{--}59 \\
\beta \quad &. \quad . \quad 71\text{--}78
\end{aligned}
$$

On now passing a long pipette charged with sulphide of ammonium to the bottom of the fluid, the sharply defined absorption-band of O-Hb appeared.

$$
\begin{aligned}
\alpha \quad &. \quad . \quad 60\text{--}65 \\
\beta \quad &. \quad . \quad 70\text{--}78.
\end{aligned}
$$

The bands were very much blacker than before, and the absorption-band in the orange had disappeared. After some time the fluid had assumed a purple tint, indicating reduction. On then examining it a single broad band, that of reduced hæmoglobin, was seen. This band extended from 65–74. The fluid in the prism was now poured into a beaker and thoroughly shaken with air, then reintroduced into the prism bottle. The colour had now changed again from purple to red, and the following reading obtained :—

$$
\begin{aligned}
\alpha \quad &. \quad . \quad 61\text{--}65 \\
\beta \quad &. \quad . \quad 70\text{--}78
\end{aligned}
$$

The two observations which I have adduced suffice to show the action which the solution of any nitrite exerts upon the colouring-matter. In my observations I have most frequently treated blood with solutions of nitrite of potassium and sodium. The action of a solution of nitrite of silver is, however, similar. The latter solution should be freshly prepared, and should not be much exposed to light. The observations should also, in this case, be made within a short time after the addition of the nitrite to the blood-solution, as otherwise the solution often becomes turbid.

The fact that ammonia altered the colour of blood which had been treated with nitrites from a chocolate-colour to red, might be explained on the supposition that some change had occurred in the constitution of the body formed under their influence, although it was quite possible that the spectrum of the body in a decidedly alkaline fluid might be different from that of the same substance in a nearly neutral solution. In order to determine whether the difference induced in the colour and spectrum was merely due to the alkalinity of the fluid, or to the presence of free ammonia, I prepared a dilute solution of ammonia. This solution contained 10 grms. of ammonia in 1 litre. I then made an exactly equivalent solution of phosphoric acid. The solutions were so exactly made that when equal quantities were mixed the fluid had no action on litmus. A weak solution of hæmoglobin was made, and a solution of nitrite of sodium added to it. As soon as the bands α and β had almost disappeared and the band in the red had appeared, I added 1 cub. centim. of the solution of ammonia. Immediately thereafter the absorption-band in the red disappeared, and the spectrum previously described as developed by the action of ammonia was seen, i. e. the orange was shaded over, and the bands α and β became more distinct.

One cub. centim. of the dilute solution of phosphoric acid was now added. Instantly the spectrum returned to the condition in which it had been before the addition of ammonia. On again adding another cub. centim. of ammonia, the band in the red disappeared and the orange became shaded over. By repeated observations I found that if care were taken to measure out exactly the same volume of the dilute acid and ammonia, the change from one spectrum to the other might be almost indefinitely repeated.

Without wishing at this stage to enter into a discussion concerning the cause of the changes which nitrites induce, I shall confine myself to remarking that the observations which I have described appeared to me to show—

First, that nitrites exert a marked influence both on the colour and the spectrum of blood, due obviously to a chemical change exerted on the colouring-matter.

Secondly, that, whatever the nature of the change, it is obviously not one which materially alters the composition of the colouring-matter, as by mere action of reducing-agents all effects of the change disappear; and

Thirdly, that the fact of reducing-agents developing, without the intervention of the atmospheric oxygen, the spectrum of oxidized hæmoglobin before exerting their reducing-action, proves that by the action of nitrites the loose oxygen of the blood is neither expelled nor removed.

III. *On the Influence of Nitrites in modifying the Respiratory Functions of the Blood.*

Under the term respiratory functions of the blood, I mean to include those physical and chemical processes which are concerned in the absorption of oxygen by blood, in its combination with the blood-colouring-matter, and its retention thereby so as to be readily available for purposes of oxidation, as well as those functions and processes which are concerned in the formation and evolution of the carbonic acid of the blood.

If the action which any substance exerts upon any or all of these various physical and chemical processes had to be ascertained, the methods of inquiry ought, in my opinion, to be equally varied.

Hitherto the method which has been almost exclusively employed to furnish evidence upon the influence which individual medicinal and poisonous substances exert upon blood, has consisted in finding out whether their addition to blood influenced the absorption of oxygen or the evolution of carbonic acid. For this purpose two samples of blood derived from the same source (one of which has been subjected to the action of the poison under examination) are brought in contact with the same quantity of oxygen or atmospheric air, and after the fluid and gas have been allowed to remain in contact for some time, the amount of oxygen absorbed and carbonic acid evolved is ascertained. This is done by determining, first, the contraction which the fluid has undergone, and, secondly, the amount of carbonic acid, nitrogen, and oxygen present in the gas after contact. Were it easy to place the blood under precisely similar circumstances in the two cases, the evidence yielded by the method would be very much greater than it really is. In order to ensure a very marked action of the blood upon the air, they must be brought together by agitation; the violence and the length of time during which this is continued are circumstances having the most marked influence upon the rapidity and amount of gaseous interchanges which take place. The difficulty of securing the same degree of agitation is very considerable, and probably introduces one of the chief errors attaching to this method of experimenting.

In such experiments I believe it to be most essential that the total volume of gas remaining after the experiment, and any contraction or dilatation which may have occurred, should be determined with the greatest accuracy. Some experimenters have not considered this essential; for in experiments to test the influence of poisonous agents on blood, they have merely made percentage analyses of the air after contact.

A second method which suggests itself as most valuable in investigating the action of poisonous agents upon the respiratory functions, is based upon the property which carbonic oxide gas possesses of displacing the oxygen of blood. The normal condition of the colouring-matter will obviously be tested in a valuable manner if we agitate it thoroughly with air and oxygen, and then bring it in contact with carbonic oxide gas. If the gas displaces as much oxygen from the poisoned blood as from normal blood which has been under exactly the same circumstances, we shall have evidence to show that the addition of the poison has, first, not prevented the normal absorption of oxygen; secondly, has not removed the oxygen from the blood by a special chemical action of its own; and thirdly, has not so altered the compound of oxygen with blood-colouring-matter as to render the oxygen irremovable by carbonic oxide gas.

A third method consists in adding the poisonous agent to blood, arterializing it as completely as possible, and then boiling out the gases of the blood *in vacuo*, determining their amount and ascertaining their composition.

It is well known that by merely exposing the blood to the temperature of the body in a Torricellian vacuum, the hæmoglobin which it contains is reduced as completely as by contact with reducing-agents. The oxygen which has been in combination with it can therefore be determined with the greatest accuracy; and in addition we can determine the amount and rate of evolution of the more or less loosely combined carbonic acid existing in the blood. This method, taken by itself, possesses perhaps more value than either of the others to which I have alluded.

The fourth method to which I must allude has already, in some cases, afforded information upon the action of certain poisons on blood which could scarcely have been observed in any other way, viz. spectrum analysis. If used in connexion with the three other methods of research which I have thought it proper to allude to in detail, we may expect that our knowledge of the mode of action of poisons on blood will become both accurate and complete. Having shown what light it appears to throw upon the action of nitrites upon blood, I shall detail the experiments which have been performed according to the three remaining methods.

Experiments in which Blood acted upon by Nitrites was agitated with Air.

All the gas-analyses made in the course of this investigation have been conducted with the admirable instrument of Dr. FRANKLAND and Mr. WARD; its value for such a research as the present one cannot be overestimated; for, besides permitting of the most accurate analysis of the gases (which have been in contact with blood) being made, it enables the experimenter to separate perfectly the gas from the blood, and to determine, with great ease and remarkable accuracy, the volume of the gas after contact with blood.

The air or gas whose action upon blood has to be studied is first measured in the eudiometer, and then passed into the laboratory-tube; into the latter is then introduced, by means of a very perfect metallic syringe of the form shown in the annexed cut, an

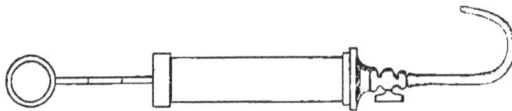

accurately measured quantity of the blood to be acted upon. The stopcocks of the eudiometer and laboratory-tube having been closed, the clamp which unites these is taken off; the laboratory-tube is then closed with the thumb, and being taken out of the mercurial trough, is agitated as long and as violently as is required. It is then allowed to remain at rest for a considerable number of hours, so as to allow the froth to subside, and (unless this has been done immediately after the agitation of the tube had been completed) it is again clamped to the eudiometer. When the time has elapsed during which it is desired to keep the gas and blood in contact, the gas can be decanted in the most perfect manner from the blood which remains in the laboratory-tube; it is then measured and analyzed. The accuracy with which this separation of gas from the blood can be effected constitutes one of the greatest merits of the instrument for such experiments. Where gases are placed in contact with blood in ordinary absorption-tubes, it is impossible to determine with sufficient accuracy the amount of gas remaining after the experiment. It is true that the gas may be decanted with a gas-pipette. It is, however, impossible in this way to decant the whole of the gas without carrying with it some of the fluid from which it is desired to separate it. The small gasometers of BUNSEN[1] which were used by Dr. HARLEY[2] in his experiments on blood, although they allow of a transference of the gas which they contain to be easily effected, do not permit the gas which they contain to be sufficiently accurately read off to enable the exact change of volume which has occurred to be ascertained.

An admirable manner of conducting these experiments consists in measuring out successively two, as nearly as possible equal, quantities of gas, which are transferred to two laboratory-tubes of the same size, each of which fits accurately to the eudiometer. Into one tube is introduced the normal blood, and into the other the same quantity of blood to which has been added the substance whose action it is desired to study. Both tubes are then agitated in the same manner and for a like time; having been allowed to remain at rest for eighteen or twenty hours, one of the laboratory-tubes is connected to the eudiometer, and its gaseous contents are decanted and measured. This laboratory-tube having been emptied of blood and decanted, the gas in the eudiometer is analyzed. The other laboratory-tube is then connected with the eudiometer, and the same series of operations gone through. If one sample of blood have been placed in contact with its measured volume of gas an hour later than the other, it will on the following day also be decanted, measured, and analyzed one hour later; during the preceding hour there has then been ample time to analyze the gas from the first tube.

I. One cubic inch of freshly defibrinated blood of the Sheep, which had been arterialized by shaking with air, was introduced in the laboratory-tube of FRANKLAND'S apparatus. Then 2 cub. centims. of a solution of nitrite of sodium of unknown strength were introduced by means of a small pipettte. A quantity of air had previously been introduced into the eudiometer and measured; this was now passed into the laboratory-tube. The mixture of blood, nitrite, and air were then thoroughly agitated by shaking; the blood readily acquired a chocolate-colour. After twenty-three hours the gas was decanted from the blood, measured, and analyzed.

[1] Gasometry, translated by Dr. ROSCOE, 1857, p. 20.
[2] "On the Influence of Physical and Chemical Agents upon Blood," Philosophical Transactions, 1865, p. 687.

Blood employed . . . . 16·38 cub. centims.
Air placed in contact with blood . . . . 225·7 vols.[1]
After contact with blood for twenty-three hours . . 225·8 ,,
After absorption of carbonic acid . . . . 225·8 vols.
After addition of hydrogen . . . . : . . 343·17 ,,
After explosion. . . . . 203·30 ,,

In 100 parts of gas.

Oxygen . . . . . 20·65
Nitrogen . . . . 79·35

The volume of gas before and after contact with blood has remained the same.

Vol. of gas before contact $=225\cdot7$ vols. $=25\cdot14$ cub. centims. at $0°$ C. and $0^{m\cdot}76$.

Oxygen present in air before contact (calculated) $=47\cdot3$ vols.

Oxygen present in air after contact (found) . $=46\cdot62$ vols.

The amount of oxygen absorbed by the blood used $= \overline{0\cdot68}$ vol. $=0\cdot075$ c. c. at $0°$ C. & $0^{m\cdot}76$.

The blood and nitrite which had been used in this experiment was at its conclusion examined with the spectroscope. The spectrum exhibited a dark absorption-band in the red, with very great faintness of the normal bands.

In the experiment just detailed, in which the blood brought in contact with air had been fully acted upon by nitrite of sodium, no marked absorption of oxygen had taken place. A cubic inch of blood had not absorbed one-tenth of a cubic centimetre of oxygen.

II. This observation is intended also to illustrate the action of nitrite of sodium upon blood.

The blood used was defibrinated blood of the Sheep, well arterialized by agitation with air. It was found by direct experiment that 1 cubic inch of this blood was fully and readily acted upon by 1·5 cub. centim. of a solution of nitrite of sodium of known strength. A sufficient quantity of air having then been introduced in the eudiometer and measured, not quite 16 cub. centims. of blood were thrown into the laboratory tube, then 2 cub. centims. of the solution of nitrite of sodium added. This quantity of solution contained 0·016 grm. of $Na\,NO_2$. The gas and blood were then agitated.

Vol. of blood about 16 cub. centims.
Air placed in contact with blood 271·2 vols.

After contact with blood and nitrite for twenty-four hours . . 271·5 vols.
After absorption of carbonic acid . . . . 268·6 ,,
After addition of hydrogen . . . . 394·03 ,,
After explosion . . . . 223·36 ,,

---

[1] I have not thought it advisable to give in the body of the paper the uncorrected readings made in the course of the gas-analyses. I have reduced all these to corrected volumes:

1 volume $=0\cdot1114$ cub. centim. at $0°$ C. and $0^{m\cdot}76$.

In 100 parts of gas.

Carbonic acid	. . . . .	1·06
Oxygen	. . . .	20·97
Nitrogen	. . . . . .	77·97
		100·00

Vol. of air before contact . =271·2 vols. =30·211 cub. centims. at 0° C. and 0ᵐ·76.

„   after contact . =271·5 „ =30·245   „   „   „

Vol. of oxygen in gas before exp. (calculated)=56·84

„   after experiment (found) . =56·89

In this case spectrum analysis proved that the blood had been fully acted upon by the nitrite. There had been no absorption of oxygen by the blood; the only changes had consisted in the diffusion of a little carbonic acid out of, and nitrogen into the blood.

III. The jugular vein of a dog was opened and the blood allowed to flow into a beaker. It was stirred with a glass rod so as to separate the fibrin, care being taken to shake it as little as possible, so as not to favour its arterialization. When the fibrin had been separated, the blood still preserved a very venous colour.

A syringeful, viz. 20 cub. centims., of this blood was introduced into a laboratory-tube, then 5 cub. centims. of a solution of nitrite of sodium were added to it. The blood and nitrite were agitated together, and allowed to remain in contact for two hours. At the end of that time air, which had been measured in the eudiometer, was transferred to the laboratory-tube and agitated with the mixture of blood and nitrite.

Immediately after the agitation the colour of the blood became much darker. After twenty hours' contact the gas was decanted, measured, and analyzed.

Volume of blood taken	20 cub. centims.
Vol. of air before contact	238·5 vols.
„ after contact	240·5 „
After absorption of carbonic acid . .	240·5 „
After addition of hydrogen	363·8 „
After explosion	224·6 „

In 100 parts of gas.

Oxygen	. . . . .	19·28
Nitrogen	. . . .	80·72
		100·00

Vol. of air before contact 238·5 vols. =43·16 cub. centims. at 0° C. and 0ᵐ·76.

Oxygen present in the gas before contact (calc.) 49·98 vols.

„   „   after contact . . 46·40 „

3·58 vols. =0·66 cub. centim. at 0° C.[1].

[1] In this case as the gas was measured at a different division of the eudiometer, the *absolute* volume as compared with the *arbitrary* volume is different from that mentioned in the note to Observation I.

From this observation it would appear that although nitrites diminish to a remarkable extent the capability which venous blood possesses of absorbing oxygen, yet they do not abolish it entirely. The action was such, however, that highly venous blood did not absorb more oxygen than average samples of thoroughly arterialized blood would have done.

IV. In this experiment, before acting upon the blood with nitrite of sodium, it was treated for half an hour with a rapid stream of carbonic acid. The object of this was partially to reduce the blood before acting upon it. The blood used was that of the Sheep, defibrinated. After passing the gas through it for half an hour, the blood, when examined in thin layers, presented a decidedly venous hue. It was then treated with nitrite-of-sodium solution. Two cub. centims. of the same solution as was used in the last experiment were added to every 25 cub. centims. of blood. Carbonic acid was then again passed through the blood.

Twenty cub. centims. were then introduced into the laboratory-tube, and a previously measured quantity of atmospheric air was passed from the eudiometer into the laboratory tube. On shaking the laboratory tube a considerable evolution of gas took place, so that when replaced into the mercurial trough a considerable quantity of mercury was expelled, and the quantity of gas in it appeared very sensibly increased.

After a contact of twenty hours the gas was decanted from the blood, measured, and analyzed.

Volume of blood taken	20 cub. centims.
Atmospheric air brought in contact with the blood	255·4 vols.
After contact with the blood for twenty hours .	322·2 ,,
After absorption of carbonic acid	254·3 ,,
After addition of hydrogen	388·96 ,,
After explosion	235·20 ,,

In 100 parts of gas.

Oxygen	15·9
Carbonic acid . . .	21·07
Nitrogen . . .	63·03
	100·00

Volume of atmospheric air taken 255·4 vols. $= 47\cdot37$ cub. centims. at 0° and $0^{m}\cdot76$
Volume of O in air taken (calculated) 53·53 vols.
Volume of O remaining . . . . 51·25 ,,
Amount of oxygen absorbed by blood$=$ 2·28 vols.$= 0\cdot421$ cub. centim. at 0° C. and $0^{m}\cdot76$.

The above experiment confirms the results of those previously described, as the amount of oxygen absorbed by blood which had been for a considerable time treated with carbonic acid was not as great as that absorbed by the same amount of the most arterialized blood. It is interesting, moreover, as furnishing an example (although an exaggerated one) of the fallacious conclusions to which mere percentage analyses of the gases which have been in contact with blood may lead. The determination of the composition of the gas would, by itself, have led to the opinion that a large quantity of oxygen had been absorbed by this very venous-looking blood. An accurate knowledge of the total amount of gas before and after contact shows that this apparent diminution of oxygen was merely relative, and due to the dilution of the gases by the large amount of carbonic acid which had, on agitation, diffused out of the fully saturated blood.

V. In the experiments now to be described, pure defibrinated and strongly arterialized blood was placed in contact with a measured volume of atmospheric air. Shortly after a like quantity of the same blood, to which nitrite of potassium had been added, was placed in contact with nearly the same volume of atmospheric air in another tube.

(a)

20 cub. centims. of thoroughly arterialized defibrinated blood of the Sheep brought in contact with a measured volume for twenty-two hours.

Vol. of air taken . . . . 337·6 vols.
After contact for twenty-two hours . 331·08 „
After absorption of carbonic acid . . 324·5 „
After addition of hydrogen . . . . 483·9 „
After explosion . . . . 295·4 „
　　　Carbonic acid . . . . 1·98
　　　Oxygen . . . . . . 18·98
　　　Nitrogen . . . . . 79·04
　　　　　　　　　　　　　100·00
Vol. of air taken . . =37·6 c. c. at 0° and 0^m·76.
O in air taken (calculated) . . 70·76 vols.
O remaining . . . . 62·84 „
　　　　　　　　　　　　 7·92 „
O absorbed by 20 cub. cen-} 0·9 c. c. at 0° C. and
tims. of arterialized blood} 0^m·76.

(b)

20 cub. centims. of the same blood as that used in (a) had 0·02 grm. of pure KNO_2 added to them, and were then brought in contact with air for twenty hours.

Vol. of air taken . . . . 348·2 vols.
After contact for twenty-two hours . 349·3 „
After absorption of carbonic acid . . 346·8 „
After addition of hydrogen . . . . 527·3 „
After explosion . . . . 319·21 „
　　　Carbonic acid . . . . 0·71
　　　Oxygen . . . . . . 19·85
　　　Nitrogen . . . . . 79·44
　　　　　　　　　　　　　100·00
Vol. of air taken . =38·8 c. c. at 0° C. and 0^m·76.
O in air taken (calculated) . . . 72·98 vols.
O remaining . . . . 69·36 „
　　　　　　　　　　　　 3·62 „
O absorbed by 20 c. c. of}
arterialized blood treat-} 0·4 c. c. at 0° C. and 0^m·76.
ed with KNO_2 . . }

VI. In the two experiments now to be described it was determined to take two portions of the same blood, to one of which a certain quantity of distilled water was added, and to the other the same quantity of water holding nitrite of potassium in solution, and to place these in tubes of the same capacity in contact with the same volume of atmospheric air, and then, in order to give the greatest possible facilities to the gas to act upon the blood, to subject them to prolonged and thorough shaking. For obvious reasons a mechanical contrivance must be devised if it be desired either to shake two tubes to the same extent or to continue the process of shaking for many hours. The instrument which I have employed for this purpose is merely a modification of a very ingenious piece of apparatus which was suggested by Mr. C. HANBURY, Jun., for washing photographic prints. It consists of a tin box, A B C D, well balanced and swinging freely in gudgeons fixed in the sides of the tin box E F, G H.   The box A B C D is divided by

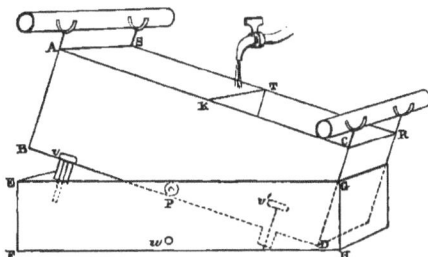

a central partition, K T, into two equal compartments.   Into the bottom of the box are
fitted, one in each side, the valves v, v', which are so constructed that when either half
of the box is dependent, the valve of that side opens.   The apparatus is brought beneath
a stream of water so that the water can enter one half of the box a little to the side of
the central partition; when a certain amount of water has accumulated, that side
of the box more than balances the other, and swings over with a powerful jerk; this
movement, however, brings the opposite side of the box under the tap, and this
commences to fill.   In the meantime the valve in the side first filled is opened and the
water is escaping.   Thus the box swings alternately from side to side.   Supports are
attached to either end of the box for supporting the absorption-tubes containing the
blood and gas.   If the apparatus be well constructed, the two tubes are placed in exactly
the same circumstances, each receiving the most thorough shake several times in a minute.

(a)

25 cub. centims. of defibrinated and well-arterialized blood of the Sheep were diluted with 10 cub.
centims. of distilled water.   25 cub. centims. of this
mixture of blood and water was placed in a tube of
the capacity of 55 cub. centims., which was immediately corked.

The tube was tied to one of the supports of the
shaking machine at 4.30 P.M. of one day, and the
process of agitation continued until 1.15 on the following day.   During this time the tubes received
1944 shakes.

After the air had been in contact (for 22 hours
and 45 minutes) with the blood, it was decanted by
means of Professor MILLER's gas-pipette and analyzed.

Vol. of gas taken . . . 154 vols.
After absorption of CO_2 . 152 „
After addition of H . . 246·8 „
After explosion . . . 155·5 „

Composition of gas in 100 parts.

CO_2 . . . . 1·29
O . . . . 19·76
N . . . . 78·95
————
100·00

(b)

The same quantity of blood as was used in (a) was
mixed with 10 cub. centims. of distilled water containing 0·1 grm. of KNO_2 in solution.   25 cub. centims. of this mixture of blood and solution of nitrite
of potassium was placed in a tube of the capacity of
55 cub. centims., which was immediately corked.

The tube b was treated in all respects like the
tube a.

After the air had been in contact with the blood for
22 hours and 45 minutes, it was decanted by means
of Professor MILLER's gas-pipette and analyzed.

Vol. of gas taken . . . 159 vols.
After absorption of CO_2 . 159 „
After addition of H . . 245·3 „
After explosion . . . 145·2 „

Composition of gas in 100 parts.

CO_2 . . . . 0·00
O . . . . 20·99
N . . . . 79·01
————
100·00

In these experiments I was only able to make percentage analyses of the gases after contact with blood, from the impossibility to measure, by the method used, the exact volume of gas after the experiment. As far as they go they fully confirm the experiments which had preceded them; as in the case of the air brought in contact with the normal blood mixed with distilled water, the oxygen was decidedly diminished, whilst in the case of that which had been in contact with the blood and nitrite, the air remained quite unaltered. I have performed several other experiments on exactly the same plan as those just described, and with precisely similar results. As, however, those which I have adduced are sufficient to illustrate the action of nitrites on blood, in so far as that action can be discovered by this mode of experimenting, I have preferred to omit giving further details, more especially as I myself attach little importance to these percentage analyses of air left in contact with blood[1].

Experiments in which Blood was acted upon by Nitrites and then treated with Carbonic Oxide.

In the experiments now to be described I have made use of carbonic oxide as a reagent for the detection and expulsion of the loosely combined oxygen existing in blood. The plan of the experiments has generally consisted in taking defibrinated blood of the Sheep and agitating it with air so as effectually to arterialize it, then treating it with a solution of some nitrite, and bringing it in contact with carbonic oxide gas. The carbonic oxide gas used was always prepared by the action of pure sulphuric acid on pure formiate of magnesium; in several of the first experiments I analyzed the gas before bringing it in contact with blood. I invariably found that the gas which I prepared in this way was absolutely pure.

As I wished in these experiments merely to find out the amount of oxygen which the carbonic oxide was capable of expelling from the blood, I confined myself to determining

---

[1] Since the experiments just described were performed, I have made others with the view of determining whether the power which the blood-colouring-matter appears to possess of ozonizing the atmospheric oxygen which comes in contact with it, is destroyed by the addition of nitrites.

ALEXANDER SCHMIDT has pointed out that when a drop of diluted blood is placed upon bibulous paper which has been soaked in tincture of guaiacum and dried, at the edges of the drop the paper assumes a blue colour such as is produced by the action of ozone. I find that the reaction can be obtained with blood which has been fully acted upon by nitrites. In order to observe this reaction, the blood should be diluted with twenty times its volume of water. I may mention that all specimens of tincture of guaiacum do not give the reaction; of three samples which I purchased, only one was found to render the paper sufficiently sensitive to show the reaction, although the other two possessed the colour, smell, and taste of the genuine tincture.

It has been shown by SCHÖNBEIN that when peroxide of hydrogen is added to blood, a copious evolution of oxygen occurs, as when the same reagent acts upon certain peroxides.

Blood which has been acted upon by nitrites effervesces on the addition of peroxide of hydrogen, just as normal blood does.

In so far, then, as we can judge by the two experiments which bear most directly on this point, we must conclude that nitrites do not destroy the ozonizing properties which the blood-colouring-matter appears to possess.

the amount of carbonic acid by absorption with potash and of oxygen by absorption with pyrogallic acid. The residual gas, consisting of carbonic oxide and nitrogen, was not further analyzed, but merely measured.

Obs. I. 25 cub. centims. of defibrinated and well-arterialized blood of the Sheep were mixed with 3 cub. centims. of a solution of KNO_2. The 3 cub. centims. contained 0·03 grm. of pure nitrite of potassium. The blood was allowed to remain in contact with the nitrite for 24 hours; at the end of that time it had acquired the characteristic chocolate-colour and spectrum which I have already described.

Pure carbonic oxide was prepared by the action of pure sulphuric acid upon chemically pure formiate of magnesium. A measured quantity of the gas was brought into the laboratory-tube, into which had already been introduced 20 cub. centims. of the mixture of blood and nitrite of potassium.

The blood and gas were agitated together and then left in contact for 24 hours. At the end of that time the gas was transferred, measured, and analyzed.

Quantity of blood used . . . .   17·8 cub. centims.
Carbonic oxide taken . . . .   273·4 vols.
After 24 hours' contact gas measured .   . . 273·0 „
After absorption of carbonic acid . . . . . 271·5 „
After absorption of oxygen . . . . .   270·5 „

.   Composition in 100 parts.

Carbonic acid . . . .   0·54
Oxygen . . . .   0·36
Carbonic oxide and nitrogen . . . 99·10
                                    ———
                                    100·00

Vol. of carbonic oxide before contact $=273·4$ vols.$=30·45$ cub. centims. at 0° and 0^{m}·76
Vol. of carbonic acid exhaled $=1·57$ vol. $=0·167$ cub. centim.   „
Vol. of oxygen exhaled $=1·0$ vol. $=0·1114$ cub. centim.   „

The mixture of blood and nitrite was examined after the completion of the experiment, and it was found that the optical characters remained unaltered. The colour was still brown, and the blood exhibited the spectrum of blood which has been acted upon with nitrites.

In the above experiment the agitation of 20 cub. centims. of a mixture containing 17·8 cub. centims. of blood (diluted to 20 cub. centims. with a solution of nitrite of potassium) with 30·45 cub. centims. of carbonic oxide gas only resulted in the evolution of 0·11 cub. centim. of oxygen gas—a quantity which is quite insignificant when compared to that invariably yielded by either venous or arterial blood. In so far as an opinion can be formed from the results of one experiment, it would then appear either that under the influence of nitrites the loose oxygen of blood-colouring-matter had been removed, or that its constitution had so altered that its oxygen was no longer capable of expulsion by carbonic oxide gas.

Obs. II. 25 cub. centims. of well-arterialized Sheep's blood were treate dwith 3 cub. centims. of distilled water containing 0·03 grm. of KNO_2 in solution, and after being mixed were allowed to remain exposed to the air ; then after an interval of two hours 20 cub. centims. of the mixture were brought in contact with a measured volume of carbonic oxide gas.

The carbonic oxide gas, which was prepared by the action of sulphuric acid upon formiate of magnesium, was found, on analysis, to be perfectly pure.

The blood was allowed to remain in contact with the gas for 21 hours, then decanted, measured, and analyzed.

Carbonic oxide taken measured	315·9 vols.
After contact with blood for 21 hours	316·2 „
After absorption with potash	314·8 „
After absorption with pyrogallic acid . . .	314·8 „

Composition in 100 parts.

Carbonic acid.	0·44
Carbonic oxide and nitrogen . .	99·56
	100·00

Carbonic oxide taken $= 315·9$ vols. $= 35·19$ cub. centims. at $0°$ C. and $0^{m}·76$
Carbonic acid exhaled $=$ 1·4 vol. $=$ 0·156   „   „   „

In the above experiment a mixture of blood and nitrite containing 17·8 cub. centims. of blood furnished on agitation with 35·19 cub. centims. of CO, no trace of oxygen gas!

Obs. III. 50 cub. centims. of defibrinated and well-arterialized blood of the Sheep were treated with a few drops of an alcoholic solution of nitrite of amyl. In a few minutes the chocolate coloration was very marked, and the spectrum had the character of that of blood treated with nitrites.

Carbonic oxide was prepared by the action of sulphuric acid on formiate of magnesium. The gas was analyzed and found to be perfectly pure. 20 cub. centims. of the blood acted upon by nitrite of amyl were brought in contact with a measured volume of it, and the gas and blood having been mixed by agitation, were allowed to remain in contact for twenty hours.

Volume of blood taken	20 cub. centims.
Carbonic oxide measured	387·9 vols.
After contact with blood and nitrite . . .	398·4 „
After absorption of carbonic acid	391·1 „
After absorption of oxygen	390·57 „

Composition of gas in 100 parts.

Oxygen	0·38
Carbonic acid	1·58
Carbonic oxide and nitrogen . .	98·04
	100·00

Gas taken (CO) measured   44·359 cub. centims. at 0° C. and 0$^{m·}$76

Carbonic acid exhaled=6·3 vols.=0·7   cub. centim.   „

Oxygen exhaled   =0·5 vol. =0·05 cub. centim.   „

The apparent increase in the volume of gas after contact with the mixture of blood and nitrite of amyl is due to the tension of the vapour of the nitrite of amyl. The amount of oxygen exhaled by the blood is quite insignificant.

Exp. IV. In the last experiment it was pointed out that the volume of gas appeared to have increased after contact with the blood; this was, however, due to the tension of the vapour of nitrite of amyl. It is a curious fact that when carbonic oxide acts upon blood, although very considerable changes take place in the composition of the gas, its volume is almost imperceptibly affected. I shall in a separate paper, devoted to the consideration of the relation which the volume of carbonic oxide absorbed by blood bears to that of the gases exhaled, point out how this insignificant change in volume can be explained. At present it will suffice to remark that, since the volume of gas does not change materially, we may bring it in contact with blood, agitate it thoroughly, and then take a fraction of the gas and analyze it without having to wait for many hours to have elapsed, as is the case in the first experiments on the action of atmospheric air upon blood. In the latter, as the gas is very liable to change in volume, we are obliged, after agitating the blood with air, to wait for a considerable number of hours so as to allow of such a subsidence of froth as will enable us perfectly to separate the gas from the fluid. In these experiments, as the volume of the gas in contact with the blood alters considerably, mere percentage analyses do not suffice to show to what extent the gas has been altered in composition, whereas in those in which atmospheric air is replaced by carbonic oxide, we may dispense with a measurement of the gas after action, and rest satisfied that, by a determination of the percentage composition, we shall obtain a full insight into the changes which have occurred. The ground upon which this assertion is based will be seen by reference to the analyses and experiments which I quote in a paper " On the Relation which the volume of CO absorbed by Blood bears to that of the O displaced."

In the experiments now to be described, recently defibrinated blood, which had been arterialized by shaking with air, was brought in contact with a measured volume of pure carbonic oxide. Having been agitated for the space of half a minute, and then left in contact with the blood for one hour and twenty minutes, the gas was decanted and the amount of oxygen which had exhaled was determined. Then the same quantity of the blood mixed with nitrite was treated in exactly the same manner, and the changes which occurred in the composition of the gas which had been added to it determined.

(a) Normal blood.

Vol. of blood used 20 cub. centims.

Vol. of CO taken 468·5 vols. =41·03 cub. centims.

Vol of gas analyzed . . 315 vols.

After absorption of CO$_2$. 313 „

After absorption of O . . 305 „

(b) Blood and Nitrite.

25 cub. centims. of the same blood as was used in (a) were treated with 5 cub. centims. of water, containing 0·05 grm. of KNO$_2$ in solution.

20 cub. centims. of the mixture brought in contact with the gas and left in contact for one hour and twenty minutes.

Composition of gas in 100 parts.

Oxygen . . . . 2·53
Carbonic acid . . . . . 0·63
Carbonic oxide and nitrogen . 96·84
——————
100·00

Assuming that the volume of gas in contact with the blood did not alter, $i. e.$ still continued after contact for one hour and twenty minutes to be 368·5 vols., or 41·63 cub. centims., the amount of oxygen exhaled by the blood would amount to 9·32 vols., or to 1·03 cub. centims. at 0° and 0ᵐ·76.

Vol. of CO taken 362·8 vols.=40·4 cub. centims.
Vol. of gas analyzed . . . 355·8 vols.
After absorption of CO_2 . . 355·0 ,,
After absorption of O . . 354·8 ,,

Composition of gas in 100 parts.

Oxygen . . . . 0·00
Carbonic acid . . . . 0·22
Carbonic oxide and nitrogen . . 99·78
——————
100·00

After the conclusion of this experiment the blood was examined with the spectroscope. The hæmoglobin bands were very faint and a marked band in the red. NH_3 caused this to disappear and shaded the orange. Sulphide of ammonium was then added. It readily reduced the blood; on blowing through it, the bands of oxidized hæmoglobin came out brilliantly. On now passing a stream of CO through the blood thus acted upon, it readily acquired the CO colour and became irreducible.

Exp. V. In the experiments now to be described, the conditions were as nearly as possible the same as in Exp. IV., with the exception that the blood used was treated with a stream of pure oxygen gas before being brought in contact with the carbonic oxide gas, or before having the nitrite of potassium added to it.

a (pure blood).

The carbonic oxide was prepared by the action of pure sulphuric acid upon pure formiate of magnesium.

Vol. of blood . . . . . 20 cub. centims.
Vol. of CO taken . . . 373·98 vols.
Period of contact with blood 60 minutes.
Gas analyzed . . . . . 340·0 vols.
After absorption of CO_2 . 340·0 ,,
After absorption of O . . 326·0 ,,

Composition of gas in 100 parts.

Oxygen . . . . 4·11
Carbonic acid . . . . . 0·00
Carbonic oxide and nitrogen . 95·89
——————
100·00

Assuming the volume of gas not to have changed during the experiment, but to have remained 373·9 vols. or 41·66 cub. centims., the O exhaled would amount to 15·37 vols.=1·71 cub. centim. at 0° C. and 0ᵐ·76.

b (blood+KNO_3).

The gas used was prepared as in a. 20 cub. centims. of a mixture of blood and solution of nitrite of potassium used. It contained
16·6 cub. centims. of blood.

Vol. of CO taken . . . 388·5 vols.
Period of contact with blood 85 minutes.
Gas analyzed . . . . . 373·0 vols.
After absorption of CO_2 . 373·0 ,,
After absorption of O . . 370·0 ,,

Composition of gas in 100 parts.

Oxygen . . . . 0·80
Carbonic acid . . . . . 0·00
Carbonic oxide and nitrogen . 99·20
——————
100·00

Assuming the volume of gas not to have changed during the experiment, but to have remained 388·5 vols. or 43·2 cub. centims., the amount of O exhaled would be 3·10 vols.=0·345 cub. centim. at 0° C. and 0ᵐ·76.

In the experiments which have preceded it has been shown that, after the action of nitrites has been exerted upon blood, carbonic oxide is incapable of thrusting out the loose oxygen; this gas, whose action on normal blood is so powerful, has for the time lost all power of action. It has also been shown that although CO loses its property of acting upon blood treated with nitrites, it acquires it again if the blood be, subsequently to the action of nitrites, acted upon by sulphide of ammonium.

It now remains for me to consider, before passing to the next series of experiments, whether nitrites have any action upon blood treated with carbonic oxide. Without detailing all the experiments which I have performed to obtain information on this point, I may state that nitrites appear to have no power of acting upon CO-blood. Whenever blood had been well agitated with carbonic oxide so as to acquire the well-known florid character and to become irreducible, it was found to be totally unaffected by the addition of solution of nitrites.

Experiments in which the Gases of the Blood are boiled out in vacuo, their amount determined, and their composition ascertained.

MAGNUS was the first experimenter who made use of a mercurial pump in order to obtain the gases of the blood[1]. Before his time several inquirers had attempted to discover the presence of dissolved gases in the blood, but their results were so contradictory that little reliance could be placed upon them. Thus whilst Sir EVERARD HOME[2] had asserted that the blood evolves gases when placed *in vacuo*, STEVENS[3] and HOFFMANN[4] in this country, and GMELIN, MITSCHERLICH, and TIEDEMANN[5] in Germany had stated that they had arrived at a different result. The failure of these experimenters was undoubtedly due to the imperfection of their air-pumps; for, as MAGNUS discovered, it is essential, in order to remove the gases which exist in a state of solution and loose chemical combination in blood, to place it in a very perfectly exhausted space. Working with his very ingenious arrangement, MAGNUS found that blood yielded *in vacuo* from about 7 to 10 volumes per cent. of mixed gases, which consisted of a mixture of carbonic acid, oxygen, and nitrogen. These numbers are extremely low when compared with those obtained by experimenters who have worked according to more recent and improved methods. After MAGNUS, LOTHAR MEYER determined the gases existing in the blood[6]. Instead of a mercurial pump, he made use of the well-known method devised by BUNSEN for boiling out the gases of water[7]. His plan of working enabled him to secure as good a vacuum as MAGNUS obtained, and to boil the blood in addition; it is probably due to this that MEYER obtained much larger quantities of gases than MAGNUS.

[1] POGGENDORFF's 'Annalen,' 1837, p. 583, "Ueber die im Blute enthaltenen Gase, Sauerstoff, Stickstoff, und Kohlensaüre."

[2] The Croonian Lecture for 1817, Philosophical Transactions, 1818, p. 181.

[3] Observations on the Blood, by W. STEVENS, London, 1832.

[4] London Medical Gazette, 1833.

[5] POGGENDORFF's 'Annalen,' xxxi. t. 289.

[6] Die Gase des Blutes, "Zeitschrift, f. rat. Med." Bd. VIII. p. 257 (1857).

[7] Gasometry, &c., by ROBERT BUNSEN. Edited by Dr. ROSCOE, 1857, p. 16.

From 100 volumes of arterial blood of the Dog he obtained from about 49·5 to 54 volumes of total gases at 0° C. and 760 millims. pressure, containing from 12·43 to 18·42 volumes of oxygen, and from 26·25 to 34·75 volumes of carbonic acid, besides from 2·83 to 5·04 volumes of nitrogen. By his method MEYER was, however, only able to get a small proportion of the carbonic acid in a free condition, by far the larger proportion only being evolved on the addition of an acid ; tartaric acid was, therefore, in his experiments, always added to the blood after the free carbonic acid had been boiled off, and then the more firmly combined carbonic acid was obtained.

The very small amount of carbonic acid which could be obtained by MEYER's method, without the addition of acid, was probably due to the blood having been mixed with from ten to twenty times its volume of water before being boiled out; for all more recent experimenters who have used the mercurial pumps for exhausting blood, and who have boiled the undiluted blood *in vacuo*, have succeeded in boiling off the greatest portion of the carbonic acid of the blood without adding any acid.

LUDWIG, and his pupils SETSCHENOW and SCHÖFFER, were the first to use the modern mercurial pump (which, as modified by GEISSLER, has proved so invaluable in researches on the blood) for the purpose of extracting the gases of the blood.

Professor PFLÜGER, with the aid of GEISSLER, of Bonn, has succeeded of late years in constructing the most perfect mercurial pumps. This physiologist has not only insisted upon the necessity of exposing the blood to a true Torricellian vacuum, as LUDWIG and his pupils had done, but contended that an arrangement whereby the gases of the blood were freed from all watery vapour was essential in order to obtain, without the addition of acids, all the carbonic acid of blood. His determinations made with the dry vacuum (das trockne vacuum) are probably the most correct which we yet possess[1]. Recently PFLÜGER has recommended that, in addition to having a drying chamber (containing sulphuric acid) in connexion with the blood-receptacle, as large a vacuum as possible should be employed, *i. e.* the vessel containing the blood to be exhausted should be connected with vessels of very large capacity, in which a Torricellian vacuum has been obtained; by this method he finds that the gases may be obtained from blood in as many minutes as the process formerly occupied hours, and that the amount of oxygen obtained is indeed greater when a large than when a small vacuum is employed.

The following Table exhibits the mean of the most recently published experiments of PFLÜGER on the gases of arterial blood.

	At 0° C. and 0ᵐ·76 pressure.	At 0° C. and 1 metre pressure.
Total gases...............	58·3 vols.	44·9 vols.
Oxygen	22·2 ,,	16·9 ,,
Carbonic acid	34·3 ,,	26·6 ,,
Nitrogen.................	1·8 ,,	1·4 ,,

---

[1] PFLÜGER, " Die Normalen Gasmengen des arteriellen Blutes nach verbesserten Methoden." Centralblatt für die Med. Wissenschaften, 26 October 1868.

In my experiments on the gases of the blood, I have used SPRENGEL'S mercurial aspirator for exhausting my apparatus and obtaining the gases of blood. The results which I have obtained with this instrument have been remarkably satisfactory, and of such a nature as to lead me to think that it may supersede the very expensive and more complicated mercurial blood-pump of GEISSLER. In the annexed diagram I have exhibited the apparatus which I employed.

J *h i k* exhibits a section of a wooden stand which is 4 feet high, and which supports three tin boxes, *x*, *y*, and *z*. The box *x* is intended to be used as a water-bath, and

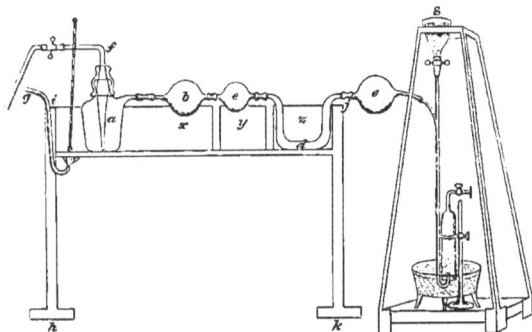

is heated by a gas-burner placed below it. *y* and *z* are tin boxes to be filled with cold water.

a exhibits the glass bottle into which is received the blood to be exhausted, and of which a larger diagram is given below. The neck *a* of this bottle is very accurately

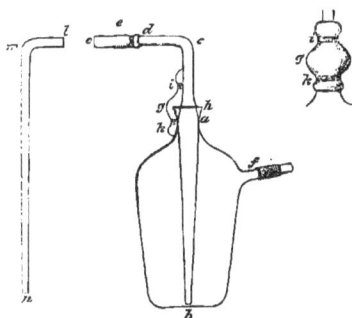

ground, and into it fits the ground tube *d c b*, which dips to the bottom of the bottle, and which is bent at *c*. The bottle has an exit-tube, *f*.

The open end of *b c d* is closed by means of a narrow tube of black india-rubber, which is wired (after having been smeared on the inside with melted india-rubber) and furnished with a very perfect steel clip. Besides having the tube *d c b* very perfectly ground, the tightness of the apparatus is further secured by pouring melted shellac into the hollow space remaining between the rim h and the tube, and thereafter slipping the caoutchouc cap g over the neck, and wiring it at i and k.

The exit-tube f is connected to a bulb b of the capacity of 100 cub. centims.; this bulb is supported in the water-bath x along with the bottle a: it is connected with the bulb c, which rests in a separate tin vessel y containing cold water; c is united to the U-tube d, and this is surrounded by cold water in the tin vessel z. The U-tube d is joined to the bulb e, which is connected to a very perfect Sprengel tube fixed in a firm wooden stand. The various tubes and bulbs are united with the aid of black tubing, melted india-rubber, and wire. Working with care there is no difficulty in getting this apparatus admirably tight. In my first experiments I made use of a more complicated Sprengel tube than the one figured in my diagram, i. e. one furnished with a gauge-tube on which a millimetre scale was etched, and which dipped into a vessel in which a very perfectly boiled barometer also dipped. In my later experiments I have used the simplest form of Sprengel tube, as I have found that the most satisfactory and perfect test consists in getting a perfectly unbroken column of mercury in the fall-tube, and observing that on allowing the mercury to flow after the apparatus has remained exhausted for some hours, not a particle of air can be removed.

When the tightness of the apparatus has been ascertained, and it is desired to use it to extract the gases of blood, the water in the bath x is heated to about 100° Fahr., and the temperature kept as constant as possible by regulating the supply of gas to the burner which heats it. A small glass tube (see diagram representing the blood-receptacle, page 613), $l\,m\,n$, having an internal diameter of a millimetre, bent at right angles at m, is filled with the blood to be analyzed. The end n is closed with the fore finger of the right hand, and with the aid of the other hand the end l of the tube is inserted into the end o of the india-rubber tube on $b\,c\,d$, until its passage further is resisted by a clip at e. The finger may now be taken away from the end n, without any risk of the blood which it contains escaping. The india-rubber tube had, however, better be wired to the glass tube $l\,m\,n$, so as to remove all risks of air getting access to it.

The blood to be analyzed is now placed in an accurately measured flask or bottle, and after its volume has been determined the vessel containing it is weighed. When this operation has been completed, the tube $l\,m\,n$, previously described, is plunged to the bottom of the flask containing the blood, and the clip e is cautiously opened; the blood is sucked into the vacuum, and when enough has entered the clip is rapidly and firmly closed. The flask which contained the blood is now weighed; after this, distilled water is allowed to flow out of an accurately graduated burette until the level of fluid in the flask is the same as it was before the blood was taken for analysis. The amount of distilled water poured out of the burette indicates the volume of the blood used.

The blood which has thus been allowed to flow into the receptacle enters into violent ebullition; a tube filled with mercury has, of course, previously been placed in the pneumatic trough connected with the Sprengel tube, and now mercury is made to flow through the latter incessantly. In about six minutes from the time when the blood was admitted into the heated blood-receptacle the evolution of gases is most rapid. Very soon, however, the bells of gas become fewer and fewer, and in about twenty-five minutes after the pumping has been commenced the process is virtually at an end. For the sake of greater accuracy I have in my experiments generally exhausted the blood for one hour.

At the end of this time the blood-receptacle and the first bulb contain a dry red mass; the second bulb contains a little fluid blood which has spirted over, whilst the U-tube contains the greater part of the water of the blood which has condensed there. Some of this is also deposited in the large bulb situated between the U-tube and the Sprengel aspirator.

After describing separately my experiments, I shall draw attention to some interesting facts connected with the use of the Sprengel tube as a blood-pump, and show that the success of my experiments throws doubt upon statements which have lately been made in reference to the conditions which are most favourable for the extraction of gases from the blood.

In order that the observations which have been made upon the gases of blood treated with nitrites should be understood, and to prove the accuracy and value of the methods used, I must quote some analyses of blood to which no nitrite was added.

I. In this experiment the gases were determined in venous blood obtained from the right side of the heart of a dog. Having exposed the external jugular vein, I passed a catheter into the right auricle, and when this had been completely filled with blood, its free extremity was inserted into the india-rubber tube attached to the glass tube leading into the blood-receiver. On opening the clip the blood flowed rapidly into the vacuous receiver.

The quantity of blood admitted was 21·52 grms. The pumping was carried on for three quarters of an hour. There was no trace of gas then coming off. At the end of that time the 21·52 grms. of blood had been, through evaporation, reduced to 8·648 grms.

The following Table exhibits the amount and composition of the gases obtained.

Gases evolved by 100 volumes of the blood of the right side of the heart of a Dog[1].

	At 0° C. and 0ᵐ·76 pressure.	At 0° C. and 1 metre pressure.
Total gases	66·03 vols.	47·59 vols.
Oxygen	12·76 „	9·19 „
Carbonic acid	49·95 „	36·00 „
Nitrogen	3·32 „	2·40 „

II. In this experiment defibrinated and well-arterialized blood of the Sheep was analyzed.

---

[1] I have thought it well to state the volume of gases obtained at 1 metre pressure, as well as at 0ᵐ·76, as the majority of the German experimenters who have lately made determinations of the gases of the blood have calculated the volumes at 0° C. and 1 metre.

The blood was allowed to flow into the receiver at one o'clock, and the exhaustion was carried on exactly for one hour.

Weight of blood used . . . . . 31 grammes.
Volume of blood used . . . . . 29·6 cub. centims.
Volumes of gas obtained . . . 167·8 vols.
After absorption of CO_2 . . . . . 70·19 ,,
After absorption of O . . . . . 6·62 ,,

100 parts of gas contain—

Oxygen . . . . 37·88
Carbonic acid . . . . . . 58·17
Nitrogen . . . . 3·95
——————
100·00

167·8 vols. of total gases   =18·69 cub. centims. at 0° C. and 0·m76.
6·62 ,,   Nitrogen   = 0·737   ,,   ,,
97·61 ,,   Carbonic acid =10·873   ,,   ,,
63·57 ,,   Oxygen   = 7·073   ,,   ,,

Gases evolved by 100 volumes of defibrinated blood of the Sheep arterialized by agitation with atmospheric air.

	At 0° C. and 0m·760 pressure.	At 0° C. and 1 metre pressure.
Total gases..............	61·12 vols.	46·97 vols.
Oxygen	23·88 ,,	18·14 ,,
Carbonic acid	36·72 ,,	27·90 ,,
Nitrogen................	2·48 ,.	1·88 ,,

III. In this experiment the same blood was used as in II., with the exception that 0·195 of pure nitrite of potassium were added to 100 parts of blood.

Weight of blood taken . . . 43·2 grammes.
Volume of blood taken . . . 41·2 cub. centims.

The bath into which the blood-receiver was plunged had a temperature of 98° Fahr. From the very moment that the blood entered the receiver the mercury was allowed to flow freely through the aspirator. The gas was collected in a laboratory-tube of FRANK-LAND's apparatus.

The evolution of gas appeared to be most brisk at 3.13, i. e. eight minutes after the blood had been admitted into the receiver. The temperature of the latter was gradually brought up to 110° Fahr. At 4.5 (after fifty-two minutes) the laboratory-tube was removed to FRANKLAND's apparatus and the gas analyzed. Still another tube having been substituted, the pumping was carried on until 4.50, the temperature of the water being taken up to 130° Fahr. Only a tiny bubble (which it would have been vain to attempt to measure) was collected during the forty-five minutes which followed the first hour's pumping.

Volume of blood exhausted 41·2 cub. centims.

Total gases obtained $\quad$ =145·8 vols.

After absorption of CO_2 = 11·65 „

After absorption of O $\quad$ = 9·58 „

Total gases obtained=145·8 vols.

Carbonic acid . . . . 134·15

Oxygen . . . . . . 2·07

Nitrogen . . . . . 9·58

Composition of gas in 100 parts.

Oxygen . . . . . . 1·41

Carbonic acid . . . . 92·00

Nitrogen . . . . . . 6·59

100 volumes of arterialized blood of the Sheep mixed with 0·195 grm. of pure nitrite of potassium yield—

	At 0° C. and 0^m·76 pressure.	At 0° C. and 1 metre pressure.
Total gases................	39·91 vols.	30·30 vols.
Oxygen	0·56 „	0·42 „
Carbonic acid	36·38 „	27·64 „
Nitrogen....................	2·97 „	2·25 „

IV. In this experiment 100 cub. centims. of recently defibrinated and thoroughly arterialized blood of the Sheep were treated with four minims of nitrite of amyl diluted with sixteen minims of rectified spirit. After having been mixed with the nitrite for one hour and forty-five minutes, a portion of the blood was admitted into the exhausted receiver. The exhaustion was carried on for one hour. After twenty-five minutes the vacuum appeared perfect, and scarcely any more bells of gas could be removed.

Weight of blood used . . . 40·5 grammes.

Volume of blood used $\quad$ . . . 38·27 cub. centims.

Total gases obtained . . . . . 178·12 vols.

After absorption of CO_2 . . 21·64 „

After absorption of O . . . . 8·64 „

Composition of gas in 100 parts.

Oxygen . . . . . . 7·29

Carbonic acid . . . . 87·85

Nitrogen . . . . . . 4·86
$$\overline{100·00}$$

Total gases obtained=178·12 vols.=19·84 cub. centims. at 0° C. and 0·^{m}76.

Oxygen $\qquad$ = 13·0 „ = 1·44 $\qquad$ „ $\qquad$ „

Carbonic acid $\quad$ =156·48 „ =17·43 $\qquad$ „ $\qquad$ „

Nitrogen $\qquad$ = 8·64 „ = 0·96 $\qquad$ „ $\qquad$ „

4 P 2

100 volumes of arterialized blood of the Sheep treated with nitrite of amyl yielded—

	At 0° C. and 0ᵐ·760 pressure.	At 0° C. and 1 metre pressure.
Total gases..............	51·84 vols.	39·39 vols.
Oxygen	3·78 ,,	2·87 ,,
Carbonic acid............	45·54 ,,	34·61 ,,
Nitrogen.................	2·52 ,,	1·91 ,,

V. In this experiment the same blood was used as in the last, but without the addition of nitrite of amyl.

Unfortunately the blood was allowed to spurt into the U-tube, and the contents of the latter beginning to froth, I discontinued the pumping before the gas had been entirely removed. As by far the larger part of the gas had been obtained, I considered that it would be amply sufficient for the purpose of comparison with IV.

$$
\begin{aligned}
&\text{Weight of blood used} \quad . \quad . \quad . \quad && 37 \text{ grammes.} \\
&\text{Volume of blood used} \quad . \quad . \quad . \quad && 34\cdot97 \text{ cub. centims.} \\
&\text{Total gases obtained} \quad . \quad . \quad . \quad . \quad && 147\cdot4 \quad \text{vols.} \\
&\text{After absorption of } CO_2 \quad . \quad . \quad . \quad && 71\cdot62 \quad ,, \\
&\text{After absorption of } O \quad . \quad . \quad . \quad && 6\cdot94 \quad ,,
\end{aligned}
$$

Composition of gas in 100 parts.

$$
\begin{aligned}
&\text{Oxygen} \quad . \quad . \quad . \quad . \quad . \quad && 43\cdot87 \\
&\text{Carbonic acid} \quad . \quad . \quad . \quad && 51\cdot41 \\
&\text{Nitrogen} \quad . \quad . \quad . \quad . \quad . \quad && 4\cdot72 \\
&&& \overline{100\cdot00}
\end{aligned}
$$

Total gases obtained 147·4   vols.=16·42 cub. centims. at 0° and 0ᵐ·760.

Oxygen	. . . .	64·68 ,,	= 7·20	,, ,,
Carbonic acid	. .	75·78 ,,	= 8·44	,, ,,
Nitrogen	. . . .	6·94 ,,	= 0·77	,, ,,

100 volumes of arterialized blood of the Sheep evolved.   (Exhaustion not completed.)

	At 0° C. and 0ᵐ·76 pressure.	At 0° C. and 1 metre pressure.
Total gases..............	46·95 vols.	35·68 vols.
Oxygen	20·58 ,,	15·64 ,,
Carbonic acid............	24·13 ,,	18·33 ,,
Nitrogen.................	2·24 ,,	1·71 ,,

VI. Five drops of nitrite of amyl were added to 50 cub. centims. of blood of the Sheep. Before the addition of the nitrite the blood had been thoroughly arterialized.

$$
\begin{aligned}
&\text{Weight of blood used} \quad . \quad . \quad . \quad && 28\cdot7 \text{ grammes.} \\
&\text{Volume of blood} \quad . \quad . \quad . \quad . \quad && 27\cdot3 \text{ cub. centims.}
\end{aligned}
$$

Exhaustion was carried on for one hour. The temperature of the bath in which the blood-receptacle was plunged was at first 100° Fahr., and was raised gradually to 120°.

The great bulk of the gas had been collected within fourteen minutes from the commencement of the exhaustion.

$$\begin{array}{ll}
\text{Volumes of gas obtained} \quad . \quad . \quad . \quad . \quad . & 83\cdot44 \\
\text{After absorption of } CO_2 . \quad . \quad . \quad . \quad . & 22\cdot2 \\
\text{After absorption of O} \quad . \quad . \quad . \quad . \quad . \quad . & 6\cdot67 \text{ vols.}
\end{array}$$

Composition of gas in 100 parts.

$$\begin{array}{ll}
\text{Oxygen} \quad . \quad . \quad . \quad . \quad . & 18\cdot61 \\
\text{Carbonic acid} \quad . & \quad . \quad 73\cdot39 \\
\text{Nitrogen} . \quad . \quad . \quad . \quad . & 8\cdot00 \\
\hline
& 100\cdot00
\end{array}$$

$83\cdot44$ vols. of total gases $= 9\cdot29$ cub. centims at $0°$ C. and $0^{\mathrm{m}}\cdot76$.

$$\begin{array}{llll}
15\cdot53 & \text{,,} \quad \text{oxygen} \quad . \quad = 1\cdot73 & \text{,,} & \text{,,} \\
61\cdot24 & \text{,,} \quad \text{carbonic acid} = 6\cdot82 & \text{,,} & \text{,,} \\
6\cdot67 & \text{,,} \quad \text{nitrogen} \quad . \quad = 0\cdot743 & \text{,,} & \text{,,}
\end{array}$$

100 volumes of blood treated with nitrite of amyl yielded—

	At 0° C. and 0ᵐ·760 pressure.	At 0° C. and 1 metre pressure.
Total gases...............	34·02 vols.	25·85 vols.
Oxygen	6·33 ,,	4·81 ,,
Carbonic acid............	24·90 ,,	18·92 ,,
Nitrogen.................	2·79 ,,	2·12 ,,

Having described the experiments in which the gases of normal blood and of blood treated with nitrites were extracted by ebullition *in vacuo*, I shall point out the facts which were either confirmed or made out in these experiments.

Three of the analyses illustrate the composition and quantity of the gases obtained from healthy blood; from these it will be seen that venous blood taken directly from the right side of the heart yielded $9\cdot19$ cub. centims. (at $0°$ C. and 1 metre pressure) of oxygen for every 100 cub. centims., that the amount obtained from two samples of arterialized blood was $18\cdot14$ and $15\cdot64$ cub. centims. (at $0°$ C. and 1 metre) for every 100 cub. centims. of blood. The addition of nitrites to well-arterialized blood was shown to lead to an enormous diminution in the amount of oxygen which could be removed by the pump; the amount of oxygen obtained being lowest in the case where the nitrite had been in contact with blood during the longest time, and highest when the nitrite was added to the blood only a short time before its exhaustion. A sample of blood which yielded $18\cdot148$ vols. of oxygen per 100 vols. of blood, after the addition of nitrite of potassium, gave up only $0\cdot425$ cub. centim. of oxygen.

In another case, where the amount and composition of the gases were determined in normal blood, and in the same blood after the addition of nitrite of amyl, it was found that 100 cub. centims. of the pure blood yielded $15\cdot64$ cub. centims. of O (at $0°$ C. and

1 metre pressure), whilst 100 cub. centims. of the same blood after the addition of nitrite of amyl yielded only 2·87 cub. centims.

These experiments, therefore, bear out, in a striking and satisfactory manner the results of those in which carbonic oxide was made to act upon blood which had been treated with nitrites.

The latter experiments showed that carbonic oxide had lost its power of thrusting out oxygen, although from certain facts which had been previously described, it appeared that the oxygen of the blood is neither expelled nor permanently taken possession of by nitrites.

The fact that the loose oxygen which can be thrust out by CO is identical with the oxygen which is given up by blood *in vacuo*, led me to suppose that bodies which, as nitrites, have the power of locking up the loose oxygen of hæmoglobin and preventing its expulsion by CO, would likewise prevent its removal by the air-pump. The hypothesis has proved to be a correct one.

I must now make a few remarks upon the use of SPRENGEL'S mercurial aspirator in the extraction of the gases of the blood.

In the detailed account of the individual experiments I have shown that, using the arrangement which I have employed, the removal of gases may be completed in a very short time (twenty-five or thirty minutes), and that the amount of gases obtained agrees admirably with the most successful of PFLÜGER'S analyses. Hitherto it has only been by PFLÜGER'S method (with the so-called dry vacuum) that the gases could be rapidly and completely removed from the blood; and as the excellence of the method has been supposed to depend upon the removal of the vapour of water by the acid in the drying chamber, it appears to be worth while examining how it is that as good results can be obtained with the arrangement which I have employed. The object of the drying chamber containing sulphuric acid, in PFLÜGER'S method, is to make the vacuum as perfect as possible by removing the steam as well as the air. It is thus intended to obviate the influence which the tension of the vapour of water would have in causing the gases to be retained. It appears to me, however, that whilst the blood is still in ebullition it would be quite hopeless to try to obtain *a dry vacuum*, as simultaneously with the drying of the air by the action of the sulphuric acid, it would again become saturated with the vapour of water given off by the blood, so that whilst the sulphuric acid probably hastened the evaporation it would scarcely influence the tension of the steam.

The sulphuric acid would, however, tend to establish a continued current of steam from the receptacle in which the blood is boiled to the chamber containing sulphuric acid, and probably this current aids very much in sweeping out the gases from the blood. When SPRENGEL'S apparatus is used, all the apparatus connected with the blood-receiver is thoroughly and rapidly swept by a continual current of steam, and to this probably some of its efficacy in removing the gases of the blood may be referred. Probably it is this current which is the cause of the superiority of PFLÜGER'S arrangement, and of my adaptation of SPRENGEL'S apparatus, over the other methods which have been employed.

On the Nature of the Action which Nitrites exert on the Colouring-matter of Blood.

Whatever might be the nature of the change induced by nitrites in hæmoglobin, it resulted from my experiments that it could not be one which deeply altered the constitution of the substance, seeing that the addition of certain reagents at once caused all the effects of the action to disappear and revealed the continued existence of oxidized hæmoglobin in the blood. Nitrites had, by my experiments, been shown to resemble in no way those agents which thrust oxygen out of the blood; on the other hand, it had been proved that the action of nitrites resulted in the locking up of the oxygen of the blood so as to render it irremoveable by CO or by a vacuum. A consideration of all the facts which I had observed led me to believe that probably nitrites might actually link themselves to oxidized hæmoglobin—a supposition which has been verified in the most ample manner.

In the experiments now to be described I have always made use of hæmoglobin prepared from the blood of the Dog, in the following manner. The dog whose blood was to be used for the preparation of hæmoglobin was kept for a considerable time under the influence of chloroform, and then bled to death.

The blood was allowed to coagulate, and the serum separated as completely as possible.

After twenty-four hours the blood-clot was broken up and firmly squeezed in linen or calico. The red fluid thus expressed was mixed with one and a half times its volume of distilled water, and set aside for three or four hours. At the end of that time the fluid was filtered through Swedish paper and mixed with one-fourth of its volume of eighty per cent. spirit. It was then placed in a vessel surrounded by ice and salt, and set aside until the following morning. The fluid was usually found to have become semi-solid from the separation of magnificent microscopic crystals of hæmoglobin. These were collected on a filter, washed with distilled water, and then dissolved in water at about 38° C. The clear red solution was filtered, treated with one-fourth of its volume of alcohol (eighty per cent.), and frozen. The hæmoglobin had at the end of twenty-four hours again crystallized out.

1. *Action of Nitrite of Potassium on Hæmoglobin.*

When solutions of hæmoglobin are treated with solutions of nitrite of potassium, the colour changes, as in the case of blood similarly acted upon, to a dark brown. Simultaneously the spectrum assumes the characters observed when nitrites act upon blood.

If a saturated solution of hæmoglobin be treated with a solution of nitrite of potassium until its colour is thoroughly changed, and if one-fourth of its volume of alcohol be then added, and the fluid set aside in a freezing-mixture, a brown deposit separates after some hours, having very much the colour of chocolate. This deposit consists of magnificent microscopic crystals, quite undistinguishable in form from those of oxidized hæmoglobin, but differing from these in appearing much less coloured, under the microscope, than the crystals of O-Hb, and in possessing only a faint yellow, but no red colour.

When a layer of these crystals is examined with the micro-spectroscope of Mr. SORBY, or with an ordinary spectroscope, the spectrum of nitrite-blood is seen to perfection.

The chocolate-coloured deposit may be dissolved in water, with the aid of a little heat, and the solution having been filtered may be treated with alcohol and frozen, when crystals identical in shape, colour, and optical properties with those precipitated the first time will again separate. The crystals when agitated with water give to it a dirty-brown colour; sometimes I have obtained them of such a size that, when shaken with water, they could be seen to float about as minute needles. If to this dirty-brown fluid a drop of ammonia be added the change is most marked and beautiful; for the fluid assumes the magnificent red colour of hæmoglobin.

If the crystals are drained by being placed on filtering paper laid on porous slabs, and then dried as rapidly as possible *in vacuo* over sulphuric acid, a reddish-brown mass is obtained which readily crumbles to powder. The crystalline form of the substance is lost in the process of drying. The dry body, when powdered, is soluble in pure water at the temperature of the body; it is much more readily soluble in water containing a trace of free ammonia; in this case a red, instead of a brown solution is obtained.

When the dry body is ignited an ash is left which has an alkaline reaction, and which contains only oxide of iron and potash.

3·379 grammes of the nitrite-of-potassium compound of hæmoglobin yielded on ignition a red ash, which was treated with water and evaporated, and then after the addition of hydrochloric acid ignited.

The chloride of potassium obtained weighed 0·014 grm.

100 grammes would therefore yield 0·47 of KCl.

5·2 grammes of the nitrite-of-potassium compound prepared at a different time yielded 0·026 grm. of KCl.

100 grammes would therefore yield 0·57 grm. of KCl.

The amount of potassium in the nitrite compound was by these analyses shown to be so small as to render very accurate results scarcely attainable. The results agree very nearly with the view that a molecule of hæmoglobin combines with a molecule of nitrite of potassium.

2. *Action of Nitrite of Sodium on Hæmoglobin.*

If a solution of nitrite of sodium be substituted for a solution of nitrite of potassium, the changes which have been described under 1 all occur, and crystals are obtained which only differ from the first by containing Na instead of K.

3. *Action of Nitrite of Silver on Hæmoglobin.*

When a solution of hæmoglobin is treated with a solution of freshly prepared and several times recrystallized nitrite of silver, it immediately becomes chocolate-coloured, and presents the spectrum of blood treated with nitrites. The solution reacts exactly as solutions of Hb which have been treated with the alkaline nitrites when ammonia and sulphide of ammonium are added. When a concentrated solution of hæmoglobin is treated with a solution of nitrite of silver and then with alcohol, and placed in a freezing-mixture, a precipitate separates, which is usually amorphous, and has a chocolate colour. Sometimes it is very readily soluble in water, holding a trace of ammonia in solution ; at other times it is soluble only with difficulty.

This precipitate, when washed and ignited, leaves a residue consisting entirely of silver and oxide of iron.

When solutions of hæmoglobin which have been treated with nitrite of silver are exposed, or set aside for some time, a precipitate separates, which is scarcely soluble in ammonia ; the solution obtained has, however, all the properties of ammoniacal solutions of the nitrite compounds of hæmoglobin.

$a.$ 4·279 grms. of the nitrite of silver compound yielded 0·0517 grm. of Ag Cl.

1·042 grm. of the same sample yielded 0·0067 grm. of $Fe_2 O_3$.

$b.$ 3·591 grms. of the nitrite of silver compound prepared on another occasion and from different blood, yielded 0·0177 grm. of Ag Cl.

The first sample, therefore, contained in 100 parts 0·759 of silver, corresponding to 1·083 of Ag NO_2 ; whilst the second contained almost exactly half the quantity, or ·346 of silver, corresponding to 0·494 grm. of Ag NO_2 per 100 parts of the compound.

These two analyses clearly indicate that, although nitrite of silver enters into actual chemical combination with hæmoglobin, the amount which is taken up varies remarkably.

PREYER has calculated the molecular weight of hæmoglobin on the assumption that each molecule of this substance yields a molecule of hæmatine. The molecular weight of hæmatine being 626 ($C_{32} H_{34} N_4 Fe O_6$), that of hæmoglobin would on the above assumption amount to 13280.

On the assumption that one molecule of hæmoglobin has the power of combining with and retaining loosely one molecule of oxygen, 13280 grammes of hæmoglobin would combine with and retain 32 grammes of oxygen, or 1 gramme of hæmoglobin would have the power of retaining 1·3 cub. centim. of oxygen at 0° C. and 1 metre pressure ; this number agrees, as perfectly as can be expected, with direct observations made to determine this point.

If one molecule of oxygenized hæmoglobin combined with one molecule of a nitrite, 13312 grammes of oxidized hæmoglobin would combine with 154 grammes of Ag NO_2, and 100 parts of the compound should contain 1·14 gramme of Ag NO_2, or 0·79 of Ag. Comparing these numbers with those which I found, we have

		Found.	
In 100 parts.	Calculated.	(1)	(2)
Silver	0·79	0·759	0·346
Iron	0·419	0·44	

4. *Action of Nitrite of Amyl on Hæmoglobin.*

If, instead of the solutions of the nitrites already mentioned, we add an alcoholic solution of nitrite of amyl to a solution of hæmoglobin, then treat the mixture with one-fourth of its volume of alcohol and expose it to cold, after some hours a brown deposit separates having very much the appearance of chocolate, and forming on microscopic examination the characters which have been already described as possessed by the bodies formed when hæmoglobin is treated with other nitrites. These crystals can, like the others, be recrystallized without undergoing any change. On one occasion (20th December 1867) I placed in a test-tube a magma of these crystals along with their mother-liquor, and on examining them now, after an interval of nearly three months, I find them still perfectly preserved and possessing all the characteristic features. This is probably due to the preservative influence of the nitrite, as I have hitherto failed in all attempts to preserve hæmoglobin crystals. I may remark that when hæmoglobin is dried, even *in vacuo*, the crystals always crumble to pieces.

In order to show that the chocolate-coloured crystals of the nitrite compounds of hæmoglobin are identical in form with those of O-hæmoglobin, I shall place two microscopic photographs at the end of this paper[1]. The first represents the compound of nitrite of potassium with hæmoglobin ; the second the body formed under the influence of nitrite of amyl, and which we may, reasoning by analogy, conclude to be a compound of nitrite of amyl with hæmoglobin. These microscopic photographs I owe to the kindness and skill of Mr. Nicol, a very able photographer in Edinburgh.

In now drawing the account of my experiments to a close, I shall state the conclusions which in my opinion may be legitimately drawn from them, and then add certain observations on the relation of the nitrite compounds of hæmoglobin to the O-, CO-, and $N_2 O_2$-compounds.

CONCLUSIONS.

1. When a solution of any nitrite acts upon the blood, peculiar changes occur in the colour, and simultaneously in the absorption-spectrum.

2. These changes in the optical properties of blood are due to the formation of compounds presenting the same crystalline form, colour, and spectrum, whatever the nitrite which has been employed in their preparation.

3. These bodies appear to be compounds of the nitrite used with oxidized hæmoglobin.

4. The substances formed by this process of chemical addition, although isomorphous with hæmoglobin, differ from it in many of those remarkable properties upon which its functions in the economy of the body depend. By this process of addition the blood-colouring-matter appears to have lost its power of absorbing oxygen.

5. The addition of nitrites to hæmoglobin appears to result in the locking up of the loosely combined oxygen, so as to make it irremoveable by CO, or by a vacuum.

OBSERVATIONS.

We have hitherto been acquainted with hæmoglobin itself as well as with its O-, CO-, and $N_2 O_2$-compounds. These compounds are all isomorphous, and possess almost the

[1] These have not been reproduced, but are preserved in the Archives of the Royal Society.

same physical characters; in all the O free hæmoglobin has apparently linked itself to a molecule of O, CO, and $N_2 O_2$ respectively, the stability of the compound being least in the case of the O- and greatest in the case of the $N_2 O_2$-compound.

All these bodies, and preeminently the O-compound, appear to be examples of a class of bodies which stand, as it were, on the boundary line which separates chemical from physical combination—to be, in fact, examples of the class of so-called molecular compounds. Like other molecular compounds their composition varies greatly within certain limits, and is influenced by circumstances and conditions which have no action on chemical compounds proper[1].

That a body possessing such a very complicated molecular structure as hæmoglobin should present numerous points of attachment, as it were, for the linking-on of such active, condensed bodies as the nitrites, is more than probable, and it is not remarkable that, as in the case of other combinations of a molecular kind, such as the union of salts with their water of crystallization, of bases with sugar, of albumen with metallic oxides, of iodine with the compound ammonias, the amount of the simpler body added to the more complex, should vary within wide limits.

Simultaneously with the observations which I have conducted, and which have shown the power of nitrites to combine with hæmoglobin, those which have lately been made by Hoppe-Seyler and Preyer, although discrepant in many particulars, seem to agree in proving that hydrocyanic acid possesses the property of linking itself to hæmoglobin, forming a body which is isomorphous with it, but which physiologically is an inert body, having lost the power which, normally, hæmoglobin seems to possess of ozonizing atmospheric oxygen.

[1] My friend Mr. James Dewar, Assistant to the Professor of Chemistry in the University of Edinburgh, first suggested to me the idea of the molecular nature of the compounds of hæmoglobin.

www.ingramcontent.com/pod-product-compliance
Lightning Source LLC
Chambersburg PA
CBHW022032190326
41519CB00010B/1687